CARLOTTA RUSOX

The Life of a Western Diamondback Rattlesnake

CARA VAN CARTER

Published by

Coyote Pack Publishing

Albuquerque, New Mexico

For permissions, inquiries, or correspondence:

info@coyotepackpublishing.com

This is a work of fiction. Any resemblance to actual persons, living or dead, real snakes, imaginary snakes, or snakes that insist they are not snakes is purely coincidental.

The animals in this book were not consulted during its writing and may disagree with certain interpretations of events.

Printed in the United States of America.

First Edition

ISBN: 978-1-972671-01-6

Cover and interior design by Coyote Pack Publishing

"If you encounter a snake in the wild, step back slowly and allow it the dignity of leaving before you start writing about it." CVC

*This book is dedicated to **Rattlesnake Solutions** for their commitment to protecting snakes and helping people coexist with wildlife. Every snake removed from a backyard and every person who learns to respect the desert shows how compassion can make room for all.*

Thank You

PROLOGUE

THE BUZZWORM

I am the buzzworm.

Mothers warn their children about me. I am the sound that makes four-legged runners startle when my tail sings beneath a rock. Those who name things have a name for me. Words in a language that tastes like nothing. But out here, names blow away. I am pattern and heat, coil and strike. I am the quiet line between sand and sky.

My desert is small and perfect. A broad sweep of creosote and cracked caliche south of the big dry wash, where the sun hammers the earth flat. In the darkness, the cold falls hard enough to slow even my blood. I was born here, slick, button-tailed. One of several that slid from my mother's cloaca into the dark of a packrat burrow. She left before the afterbirth cooled. That is the way of things. No teaching. No guarding. Just her scent fading and the desert's lesson: nothing here is given. You take, or you are taken.

I have shed more skins than I can remember, each a little longer, each scraped off against stone or root or the rough burrow wall. The newest sits behind a rock, a ghost of who I was. Every rattle segment is a shedding I survived. Hawks, kingsnakes, the fast bare ground, the swinging sticks with flat hard things. I carry their marks under these diamonds. In the still hours, I can taste the blood from when the two-legger pinned me to the ground, and I left half my stripes on the desert floor. The rattle regrew, shorter, crooked, louder. Good. A crooked song still warns.

The two-leggers think the rattle means anger. It does not. It is what I do last, before all else. I freeze. I coil. I give them the warning sound. Most hear it. Most back away, their heavy, hard skin covers fading. Those who ignore it, who keep closing in, get what the sound threatens. The desert tallies its own debts, and I do not balance them. I add only when needed.

I have no tribe. No, mate, that stayed past the act. No, young, I watched beyond the time they left the burrows where I gave birth. I have only this area of ground, the warmth of flat rocks in the new heat, and the quiet movements of a kangaroo rat under the moon, its body a warm shape felt against the cool soil, detected by my face before my eyes. That is enough. More than enough.

You will read my sheds. Birth in the dark. First skin peeling away. The first strike that hit nothing, the first strike that did. The long cold when the blood goes slow, the body barely moves, and other bodies press close for warmth. The shadow from above. The scent that pulled me across open ground in the heat. I followed something I had never smelled before, to a male I did not stay with. The weight of the young inside me. The strange heaviness of my own body turned against itself. The litter I left was left the way I was left. The flat, hard thing that came down from above. It took part of my rattle, part of my speed. What followed came slower, less accurate. Prey got away more often than it used to.

Read them as you will. I did not arrange them for you.

I do not speak for all snakes. I speak for my own body, for

the ground I rest on, and for the rattle I carry, one that has been broken, has grown back, has broken again, and yet still produces sound when I need it. The desert tested me as the sun tests stone: by pushing until the stone cracks or holds.

I held.

What did not kill me did not make me wiser, stronger, or better. It simply made my next strike cleaner and the next wait longer.

Listen when the tail sings.

Step wide, or come closer and learn what lives in these quiet coils.

I am Carlotta Rusox, daughter of no one. The mother of none that I kept.

This is the only testament I will ever give.

CHAPTER ONE

BORN IN THE DARK

It is pressure.

Something behind me, pushing. Wet walls sliding against skin I did not know I had. All I know is a tunnel, and the tunnel is closing, and I am moving through it whether I want to or not.

Then open. Air. Warm. Heavy. Thick with water smell and the musk of old fur and dry sticks. I spill onto damp earth, and the earth is soft under me, and everything is vibration.

I am not alone. Bodies beside me, on top of me, sliding past. Small and soft and hot. A pile of us. Each slimy, squirming thing has a nub at the end of its tail that clicks faintly against the others when we shift. My own nub taps the dirt floor once, and the sound disappears into the dark.

We are in a burrow. Packrat. The smell is layered deep: old urine, cactus fiber, the bones of things eaten long ago, and over it all the sharp green bite of creosote soaked by rain. The rain is outside, but its sound fills the space like another pulse, steady and low, and the ground hums with it.

She is here. The large adult. Her body stretches across the burrow floor, and the warmth coming from her is the strongest thing in the darkness. I cannot see her, but I feel her along my side, the steady movement of her breathing, the heaviness of muscles beneath scales that are thicker and harder than anything I have touched. Her scent overwhelms the burrow. It is what covers everything else, and it is the scent that I also carry, almost the same.

My tongue moves. I did not tell it to. It slips out, touches the air, pulls back. Something arrives that is neither sight nor sound: wet, warm, blood close, safe. I do it again. Wet, warm, blood close, safe. The tongue knows things the rest of me has not caught up to.

The large one shifts. Her coils rearrange, and the burrow's heat shifts. Small bodies pressed against my side are pushed away. A few of them make sounds like air escaping a crack in stone.

I feel hunger. This hunger has always been present, older than the air or the first flick of my tongue. It sits behind my eyes, like another set of eyes searching for something I haven't yet learned to find.

Thunder moves through the ground. Not the rain. Deeper. It rolls from somewhere far off, and the burrow walls carry it through my belly scales and up through the bones of my skull. Everything in the burrow goes still when it passes. Then the small bodies start moving again, and so do I.

The darkness holds. I do not know how long. There is no light in the burrow except what leaks from the entrance, and even that is gray and weak and smells of water. The rain changes speed. Faster, then slower, then faster again. The ground hums differently with each change.

She moves.

The large creature uncoils from around us in a long pull. Her body slides across mine, across the others. The heat she leaves behind begins to cool immediately. Her tail passes over my head. I taste the air behind her; her scent already fading.

She moves toward the entrance. The gray light catches the sharp edge of her scales for a moment. Diamonds and dust. Then she slips through the opening and is gone.

The warmth she held in the burrow begins to leak out after her.

I wait for her to come back. The scent trail she left points toward the entrance, and I taste it in short intervals, expecting it to freshen. It does not. It fades. Each tongue flick finds less of her until what remains is only the old smell baked into the walls, the smell that was here before we were.

She does not come back.

The others do not seem to notice, or they notice, and it changes nothing. Some of them crawl over me, heading for the entrance. Then another. Their heat trails cross and tangle on the burrow floor.

I am pressed against the back wall. The dirt is cooler here. I can feel the stone beneath the dirt, and beyond the stone, the deep cold of earth that the sun never reaches. My belly scales read the ground the way my tongue reads the air. Temperature, texture, moisture, slope. Everything is information, and none of it has a use yet.

The hunger sharpens.

I move toward the entrance. The tunnel narrows, and my body finds its way through by pressing against the sides, reading the width with my ribs. Packrat sticks scratch along my back. A cactus spine, old and dry, catches a scale and pulls free.

The entrance is a hole of gray light and moving air. The rain has slowed. Water drips from the lip of the burrow and hits the ground below in a rhythm I can feel through the stone.

I push my head out into the open.

The desert is enormous.

The sky presses down from every direction. No end. The

air grazes my scales, cooler than the burrow. It carries scents layered so thick I cannot separate them: creosote, wet stone, something sharp and green, perhaps torn plant flesh. Something warm and fleeting vanishes before my tongue can catch it.

The ground is different. Not the soft packed dirt of the burrow, but caliche, cracked and hard, with gravel between the cracks. My belly scales grip it differently. Each push forward takes more effort. The rain has left puddles in the low spots, and when I cross, the water is cold against my underside and tastes of minerals and dust.

A body from the burrow is ahead of me. I can feel the faint trace of its scent on the wet ground, fading. It has gone left, toward a creosote bush. I do not follow it. There is no reason to follow it and no reason not to. It is just another body now, getting farther away.

The temperature is dropping. The rain has stopped, but the clouds hold the heat down, and the ground is cooling beneath me. I can feel my body slowing. The cold enters through the belly scales, then the muscles, then the hinges of the jaw. My tongue flicks come slower.

I need warmth.

I move along the base of a creosote bush, tasting the air, tasting the dirt. The roots have pushed up small ridges in the caliche, and in the spaces between the ridges, the earth is fractionally warmer. I press into some of these spaces. My body curls without my telling it to, a tight coil that puts the most mass at the center and leaves the least surface open to the air.

The button on my tail is trapped under the lowest coil. It clicks once against the root and goes quiet.

Dark comes. Or what I will later learn to call dark. The heat drains out of everything. The ground, the air, the creosote above me. My blood slows to match it. The hunger is still there, but it is quieter now, pushed down by the cold into something I can carry without moving.

I hold still. There is nothing else to do. The desert is working on me the same way it works on the rocks and the dirt and the roots of the bush above me. Pressing the heat out. Waiting to see what is left when the cold has taken everything it can.

I am left.

The beginning comes as heat. Not light. Heat. The ground warms under my belly before the sky changes color. The warmth enters through the scales, reaches the muscles, loosens them, and the tongue starts working again.

I taste the air. Dry now. The rain smell is already fading, replaced by dust and the baked chemical scent of creosote leaves in the sun. Something else. Something small and warm, moving low to the ground, close. Not in the burrow. Out in the open, a few of my body lengths away, maybe more.

My head turns toward it. The pits on my face read its heat against the cool ground. A small shape, quick, low. Lizard. It stops on a flat stone in the new sun and holds still.

The hunger tightens.

I do not move. Not because I have decided not to. Because my body is still cold, the muscles are still slow, and the distance is too far. The lizard sits on its stone. I sit in my coil. The sun works on both of us.

The lizard moves. It crosses from its stone to another stone, closer to the creosote, closer to me. Then it is gone, fast, into a crack I could not reach.

The hunger stays.

I uncoil slowly and move out from under the root. The caliche is warming. My belly scales read the temperature rising by increments as the sun climbs. I move south along the base of the bush, tasting the ground, tasting the air. More creosote. More dust. The faint ghost of a kangaroo rat trail, not old, already dry.

I find a shallow depression where a couple of rocks lean against each other, and the gap between them is just wide enough. I slide in. The stone is cool on my back and warm on my belly where it faces the sun. I coil again. The button clicks once against the rock.

The birth burrow is behind me now. The scent of it, the scent of the others, the scent of the large one who left, all of it is gone from the air. There is only the desert. Creosote and caliche and stone and dust and the things that move through it that I cannot yet catch.

The hunger sits behind my eyes.

I wait with it.

CHAPTER TWO

FIRST SKIN

Something is wrong with my body.

It starts after the hunger has come and gone and come again. The scales along my sides feel tight. Not the tightness of cold, which I have already learned, but something underneath the scales themselves, a pressure pushing out from inside. My skin does not fit the way it did when I left the burrow.

The tightness gets worse. When the sun has crossed overhead, my eyes have changed. A film slides over them, occluded and thick, and the desert goes soft at the edges. I can still feel the heat. I can still taste the air. But the shapes that were sharp before are now smeared, and the light that was clear is now gray and flat, no matter how high the sun climbs.

I stop moving.

This is not a decision. My body stops. The tightness and the clouded eyes and the strange dull feeling along every scale, all of it says the same thing: hold still. Find cover. Wait.

I am wedged between the rocks where I spent the last darkness. The gap is narrow enough that nothing large can reach me from above and nothing fast can come at me from the sides. The stone holds the early sun to the face and stays cool on the other side. I press my belly against the warm side and coil as tight as the space allows.

The hunger is there. It has been there since the burrow. But it is quieter now, pushed behind whatever is happening to my skin. The skin is the louder thing.

The film on my eyes thickens. I cannot see the creosote bush even at a few body lengths. I cannot see the ground past the edge of the rocks. The desert is heat and scent and vibration only, which is most of what it has always been, but the missing part matters more than I knew.

I taste the air constantly. Creosote. Dust. The faint chemical trail of a beetle crossing caliche somewhere nearby. A kangaroo rat passed this way in the dark. Its scent is layered in the dust between my rocks, already stale, already cooling in the heat. I cannot hunt it. I cannot hunt anything. My body will not uncoil. The tightness holds me in place the way the cold did, but this is not cold. This is something building.

In the next light, still with clouded eyes, the skin at the tip of my nose loosens.

It starts as a small thing. A flap of scale at the very front of my face, where my snout meets the air, peels back the way a dry leaf curls from a branch. I push my nose against the rough edge of the rock beside me. The flap catches. I push harder. More of it separates, rolling back along the top of my head like a sheath being pulled off a root.

The relief is immediate. Where the old skin comes away, the new skin beneath it is smooth and bright, and the air touches it differently. Sharper. Cleaner. Every grain of dust and every molecule of scent lands harder on the new surface.

I keep pushing. The rock is rough enough. I press the side of my head against it and pull forward, and the old skin rolls back farther, past the place where my eyes sit beneath the film. The eye caps hold for a moment, stretched tight, and then they peel away with the rest, and the desert comes back.

Not gradually. All at once. The creosote bush is there, its branches sharp against the sky. The caliche is there, cracked and white in the sun. A whiptail lizard is frozen on a flat stone several body lengths away, its ribs moving fast with breathing. I can see the individual grains of sand on the rock face beside me.

The old skin bunches behind my head now, a crumpled tube that still clings to my body from the neck back. I move forward. The rock edges catch the old skin, and it slides off me as I go, inside out, peeling away from my sides, my belly, the long muscles of my back. I can feel each inch of it separating, the dull pull of the old surface releasing the new beneath it.

I push through the gap between the rocks, and the last of the old skin catches on a sharp edge and holds. My tail slides free. The button at the tip pulls through the old skin's tail sheath, and behind it, something new has formed. A segment. Hollow, dry, lighter than the button. When my tail moves against the rock, it makes a different sound. Not the flat click of the button alone. A small, faint buzz. Barely there. The whisper of a rattle.

The old skin hangs on the rock behind me like a ghost. Transparent, paper thin, every scale printed on it. I do not look back at it. There is nothing in it that I need.

The desert after shedding is louder.

That is the wrong word. There are no words. But everything that touches the new skin registers more. The caliche under my belly is textured in ways I could not read before. The air on my scales carries temperature gradients I missed. The sun is hotter on the new surface, or the new surface feels the sun more. My tongue flicks bring back sharper information: the whiptail lizard is still there, its scent clean and distinct against the rock dust. Male. Young. Fast.

Too fast. The hunger notes it and lets it go.

I move away from the rocks after several darks and lights. The new skin slides over caliche and gravel, and the belly scales grip differently now, the edges sharper, the contact fuller. I can feel the slope of the ground more clearly. I can feel where the soil is compacted, where it is loose, and where a burrow entrance has softened the surface.

The creosote bush where I sheltered first is behind me now. I move south and east, following the contour of a shallow wash. The wash is dry. The rain that soaked my first dark has already been absorbed by the ground and pulled back into the sky. The caliche at the bottom of the wash is cracked into plates, and the cracks hold the scent of things that passed through when the water was running: a toad, heavy and slow. A packrat, dragging something. The old trail of a snake much larger than me, its belly scent wide and deep in the dried mud.

I cross the wash and climb the low bank on the other side. The ground here is different. Rockier. More gaps between stones. More places where a body my size can fit, and a body larger than mine cannot. I find a space beneath a flat rock that tilts south and holds the sun on its surface. The heat radiates down through the stone, and the air beneath it is warm and still, smelling of dry earth and nothing else.

I coil under the rock. The new rattle segment taps the stone once, and the sound is thin but real. Something I did not have before. Something that is mine.

The hunger has been waiting. It comes forward again now that the shed is done, the body is moving, and the new skin is working. The pull behind the eyes sharpens. The tongue flicks come faster, reading the air in every direction, searching for the thing I have not yet found: a warm body small enough and slow enough and close enough that my strike can reach it before it is gone.

The whiptail lizard is long behind me. But there will be others. The desert is full of small, warm things that move along the ground, stop in patches of sun, and do not always watch for what waits beneath the nearest rock.

I am beneath the nearest rock.

The new skin tightens around me as I coil, not the tightness of the old skin that needed shedding but the tightness of a body that fits itself exactly, every scale seated, every muscle aligned, the whole length of me drawn into a circle with the head at the center and the rattle at the edge.

I wait.

The desert does what it always does. It heats up. It cools down. Things move through it. Some of them will come close enough.

The hunger and I wait.

CHAPTER THREE

FIRST STRIKE

The lizard appears at first light, when the ground is still cool enough to slow my body.

I have been under the flat rock for several light and darks. The hunger has changed. It is not the dull pull it was. Now it is sharper. A tightness in the muscles behind the jaw. A readiness in the coil that was not there before. My body wants to move in a way it has not wanted to before. Not forward. Not away. Toward.

The lizard is a side blotched. Small, fast, brown against brown. It comes out of a crack in the caliche a few body lengths from my rock and sits in the new heat. Its ribs pump fast. Its body heat is faint against the warming ground. But my face reads the difference. A spot of warmth, slightly above the surface temperature. It pulses with the rhythm of its blood.

I watch it. My eyes are clear from the shed, the shape sharp. But the pits on my face are what hold it, reading its temperature against the cooler ground. The lizard is a warm shape against a cooler background. The pits track it the way my tongue tracks scent: without effort, without decision. Information arrives. The body responds.

I uncoil. Slowly. Loop loosening, then another. My belly scales flatten against the ground, silent. The new skin grips the stone without sound. I align my body along the underside of the rock, head aimed at the lizard, and wait.

It does not see me. It is watching the sky and the open ground. Its head turns in small jerks, checking for movement, checking for shadows from above. It does not check the dark space under the rock a few body lengths to its left.

I am close enough. I think I am close enough. The muscles behind my head are loaded. The coil in my body is not the resting coil. It is something else. Tighter, pulled back like a dry branch bent against a stone. The venom glands behind my eyes feel full, heavy. There is a pressure in my fangs I have not felt before. A readiness to open, to swing forward, to puncture.

The lizard shifts on its stone. Now, closer to the crack it came from. Farther from me.

I strike.

The motion is faster than anything I have done before. My head launches forward. My mouth opens. The fangs swing down. The distance between the lizard and me disappears. I feel the impact in my jaw. I feel the fangs hit something. I feel the venom glands compress.

I feel the fangs scrape across dry stone and close on nothing.

The lizard is gone. It was never where my strike landed. It moved the instant my body moved. Maybe even before, with some ground vibration or a shift in the air that told it what was coming a breath before it arrived. It is in the crack now. I can taste its scent on the stone where it sat, warm, alive, fast. The scent is rich, close, useless.

The hunger does not ease. The muscles reset. The coil rebuilds.

I return to the space under the rock and wait.

The sun crosses, goes dark, and recrosses several times. I taste and re-taste the air. Kangaroo rat trails in the dust. A scorpion moving along the base of a stone in the dark, its body barely warmer than the ground, not worth the strike. A beetle, slow and hard shelled, crosses the open caliche. I let it pass.

After a while, a different lizard appears. Bigger than the side blotched. A banded gecko, pale and soft bodied, moves along the shaded side of a rock in the last heat. It is slower. Its scent is different: muskier, wetter. Its body throws more heat into the cooling ground, and my pits read it clearly: a bright, warm shape moving left to right across my field.

I do not strike from the same distance. I wait. The gecko moves closer. It pauses at the edge of my rock, on the stone, its belly low to the ground. I can feel its weight through the rock's vibration. Small, but there.

Closer.

It steps off the rock onto the caliche and turns toward the gap where I am coiled. It does not see me. The light is behind it, and I am in shadow, and my pattern against the dark stone looks like more stone.

I strike.

This time, the fangs connect. I feel them punch through the skin. The venom glands compress and release. I feel the hot flood of fluid leaving my body and entering another body. The gecko jerks hard. Its tail whips against my head. I open my mouth and let go.

The gecko runs.

This is the part I did not know. The prey does not fall. It does not stop. It runs, dragging itself across the caliche in a broken line. For a moment, hunger screams at me. Chase it. Grab it again. Hold it, crush it, and swallow it whole.

I do not chase it. My body does not chase. The coil resets. My head pulls back to center. My tongue flicks out, and I feel my focus shift; sight fades; smell dominates. The air where the gecko was is dense with information: blood, venom, musk, fear chemicals, the directional trail of a body moving away to the left.

I wait.

The waiting is hard. The hunger is louder than it has been since the burrow. The gecko's scent is everywhere on the ground in front of me. A trail of blood and leaking venom points exactly where it went. But the body says wait. Something older than hunger, something in the coil itself, says hold still and let the venom work.

I wait.

The sun drops lower. The shadows of the creosote bushes stretch long across the caliche. The gecko's tail no longer moves. The scent is still there, still strong, but the movement at the end of it has stopped.

I uncoil and track the trail.

My tongue leads. It touches the ground, sensation sharp and cool, reading the chemical path the gecko left. Blood here: metallic and thick. Venom here: bitter, burning the tip. I sense where it stumbled, belly dragging; where it turned, then collapsed against the base of a small stone. The trail is maybe a few body lengths long. Not far. But the waiting stretched it.

The gecko is on its side against the stone. I shift focus, heat sense yields to scent and touch. Its body is still warm. The ribs are not moving. The musk smell is heavy, and the venom smell is mixed with it, and there is a sweetness under both that I will later learn to recognize as the scent of something that has stopped being alive.

I approach from the head end. My jaws gape wider than I imagined. My lower jaw unhinges. Each side slides independently. My mouth yawns into a space larger than my head. I clamp the gecko's head in my mouth. Its body yields. I swallow slowly, muscles rippling as the gecko slides backward into my throat in waves.

It drags on. The gecko dwarfs anything that has entered me. It is the first thing that has entered me. My body clamps around it. I track its shape in my throat, then in my belly. A hard lump punches against my coils from within. My jaw snaps back. My mouth seals. I taste the air, thick with gecko, blood, venom, and my own body heat.

The hunger eases.

Not gone. Changed. The tight pull behind the eyes softens, and the muscles in the coil relax, and the urgency that has been driving every tongue flick goes quiet. There is a heaviness now. A fullness. My body wants to find a warm place and hold still and let the heat and the chemicals inside me do whatever they are going to do to the thing I just swallowed.

I find a flat stone in the last of the sun and coil on its surface. The stone holds the heat against my belly, and the gecko inside me begins to dissolve. I can feel it changing, softening, becoming something my body can use.

The rattle segment taps the stone once as I settle. That small thin buzz. I am longer than I was when I left the burrow. I am heavier by a gecko. The desert gave me nothing. I took it.

The sun drops behind the ridge. The stone cools. I stay on it until the warmth is gone, then slide into a crack at its base and coil in the dark.

The hunger will come back. It always comes back. But for now, the body is full, and the muscles are loose, and the new skin fits, and the ground beneath me holds enough warmth to last until the sun returns.

I sleep the way stones sleep. Still, and close to the earth, and ready for the sun.

CHAPTER FOUR

ABOVE

I learn about the sky the way I learn everything else. Something tries to kill me, and I survive it, and after that, I know.

Early in the light is bright and dry. The rains are gone now, behind me, and the desert has returned to what I will learn is its usual state: heat and dust and a sky that presses down and hard from horizon to horizon. I am moving along the edge of a wash, following the base of a low bank where mesquite roots have cracked the caliche, opening gaps in the soil. It is good hunting ground. The gaps hold lizards. The lizard trails are fresh, laid down in the cool before the heat returns, and my tongue reads them clearly against the warming ground.

I am bigger now. I have had a few sheds. The rattle has added segments, and when I move across loose stone, it makes a faint, dry sound that is almost a buzz. Almost. Not yet the full voice it will become. My strike is faster than it used to be. My aim is better. I have had a few kills since the gecko, both lizards, both taken the same way: strike, release, wait, follow the trail. The hunger comes and goes in longer intervals now. I am learning its rhythm the way I learned the rhythm of heat and cold, sun and dark. It builds, I hunt, I eat, it quiets. Then it builds again.

I am out in the open when it happens.

The wash bank is low here, barely shaded along its base, and I have moved past the last mesquite into a stretch of bare caliche dotted with small stones and dry grass stalks. I am following a lizard trail that runs along the surface toward a cluster of prickly pear many body lengths ahead. The sun is directly above. My shadow is small and tight beneath me. The ground is hot under my belly scales, almost too hot. I am moving fast to cross the open stretch and reach the shade of the prickly pear.

I do not see the shadow because my shadow is already beneath me, and the new shadow comes from behind, sliding across the ground ahead of me, growing. By the time I register that the shadow is not mine, that it is moving faster than I am moving, that it is getting larger instead of smaller, the air changes.

A pressure wave hits me from above. Not wind. Displacement. Something big and fast is dropping through the air, pushing the air ahead of it, and the pushed air hits my scales like a rock pressing down on my back.

I freeze.

This is not a decision either. My body stops the way it stopped during the shed. Every muscle locks. The coil flattens. My head drops to the ground. My pattern, the diamonds and dust that run the length of my body, press into the caliche, and I become a strip of ground that was not there a moment ago.

The shadow passes over me. Close. I can feel the turbulence of it on the scales along my spine. A heat source, large and fast, the body temperature of a warm blooded thing much bigger than any lizard, drives through the air less than a body length above me, and the wind of its passage rolls me half a turn against the ground.

Then it is past.

I hold still. The shadow swings in a wide arc ahead of me, curves, and starts to come back. I can track it by the way it blocks the sun: the ground in front of me goes dark, then light, then dark again. The thing is circling.

It lands.

Not on me. On the open caliche, fewer body lengths ahead, near the prickly pear I was heading for. The impact comes through the ground, a heavy double strike of something hard and sharp hitting stone. Then scraping. Then nothing.

I cannot see it well from this distance and at this low angle. A shape, upright and large, blocking the sky where it sits. Dark wings folded. A head that moves in sharp jerks, turning, scanning. The heat of it is strong even from here, a dense warm mass that my pits read against the cool shadow it throws behind itself.

It is looking for me.

I do not know this the way a two-legger would. I know it because the head keeps turning toward me, the body stays oriented toward the stretch of open ground I am pressed against, and the thing does not fly again. It is waiting the way I wait when I have struck something and let it go. Still. Fixed on the spot where I last moved.

I do not move.

The sun beats on my back. The caliche is burning under my belly now, the heat building past the point where my body can use it and into the range where it starts to cost me. My muscles want to shift, to move to shade, to find cooler ground. I hold them still. The thing on the ground ahead of me is larger than anything my strike was built to reach. and it is built for speed, and it has talons that I felt in the wind of its initial pass, sharp and hard and designed for exactly my diameter.

It waits. I wait. The sun moves.

A sound comes from it. High, thin, a short repeating cry that carries across the flat ground and echoes off the wash bank behind me. It calls rapidly. Then it opens its wings, and the wingspan is enormous, wider than I am long, and the air displacement hits me again as it lifts. The shadow sweeps over me again. I feel the talons in the wind. They pass over my back close enough that the disturbed air presses my scales flat against the caliche.

Then it climbs. The shadow shrinks. The wingbeats grow fainter, each a pulse of displaced air that I feel less and less as the distance opens. The heat signature drifts north, dims, and is gone.

I wait.

I will know this as hawk.

I wait longer than I need to. The sun moves across the sky long before I uncurl my muscles and lift my head from the ground. The caliche where my belly pressed has a faint outline of my body in the dust, the pattern of scales printed in dirt. I taste the air. The thing's scent is still there, faint, a sharp, dry smell of feathers and meat and something acidic that I will learn to associate with the droppings of things that eat from above.

I do not cross the open ground to the prickly pear. I turn back. I follow the wash bank south, staying in the thin strip of shade at its base, moving slowly, stopping at every sound from above. A creosote bush offers deeper cover, and I slide beneath it and coil in the roots and press my back against the branches and wait for the sky to empty.

The sky does not empty. It never empties. This is the thing I did not know, and now I do. The ground has its dangers: the fast things, the big things, the cold. But the sky has its own, and the sky's dangers come from where I cannot strike, cannot rattle, cannot coil against. They come from above, from the direction my body cannot face, and they are faster than anything on the ground.

From that strike, I read the sky differently. Not with my eyes, which look forward and down. With my skin. The pressure changes when something drops. The shadow that moves against the direction of the heat. The displacement of air that arrives before the talons.

I stay under cover as much as I can when I cross open ground. I move along the bases of things: bushes, rocks, banks, the roots of paloverde trees that arch over the wash. I wait for my crossings when the sky is still. I freeze when a shadow passes that is not a cloud.

The hawk does not come back. Other hawks will. Other shadows, other talons, other pressure waves from above. The sky is not empty, and it is not safe, and nothing in it is too small to notice me.

The rattle on my tail is useless against the sky. It warns about things on the ground. It warns things that hear and understand that the sound means stop, go back, this is not worth it. The things above do not hear it until they are already falling, and by then, the only defense is to not be where they expect me to be.

I learn this with my body. I do not think it. I do not conclude it. I live inside it, a flinch in every muscle when a shadow crosses the sun, an automatic flattening that happens before I know it.

The ground is mine. The sky belongs to something else.

I stay low.

CHAPTER FIVE

LONG COLD

The heat leaves slowly and then all at once.

The heat has been shortening. The sun crosses a lower arc, and the shadows it throws are longer and cooler, and the ground does not hold warmth the way it did. I can feel the change in my belly scales. The caliche that used to burn at high heat now only warms. The rocks that radiated heat deep into the dark now cool before the stars come out.

My body responds before I understand what it is responding to. The hunger fades. Not the sharp fade of a fresh kill quieting the pull behind my eyes, but a slow general dulling, as if the pull that drives me toward prey is winding down on its own. I hunt less. I miss a lizard I would have taken easily before, and do not bother to track the scent trail. I eat a small kangaroo rat at the beginning of the cooling, and after that, the hunger goes so quiet that I can barely feel it.

I am slower. When the light comes, it takes longer to warm me, and it does not build as much heat as it used to. I bask more. I lay on a south facing rock with my body stretched to its full length, absorbing every warmth the sun has to offer, and when the sun drops behind the ridge, I am already cooling, already stiffening, already sliding toward the heavy stillness that comes when the blood goes thick with cold.

I need to go underground.

I search. I follow the base of the wash south, tasting the air for the scent of other snakes. I find it: a trail laid into the dust along the edge of a rock outcrop, thick and layered, the musk of multiple bodies passing the same way over multiple crossings. The trail is old on top and older beneath, and the oldest layers have sunk into the soil itself, a chemical record of bodies that passed this way before I was born.

I follow it.

The outcrop rises from the caliche in a jumble of cracked stone and packed earth. The south face catches the low sun, and the rock there is warmer than the surrounding ground. At the base of the south face, where slabs lean against each other, there is an opening. A crack, barely wider than my body, angling down into the dark. The scent pouring out of it is heavy: snake musk, so many individuals, layered, dense, and warm. Air rises from the crack, carrying the heat of bodies packed together underground.

I slide in.

The crack narrows, then widens into a gap between the rock slabs. The floor is smooth, polished by the passage of scales over stone. The temperature rises as I descend. Not warm. Not surface heat. But warmer than the surface, warmer than the air, a steady underground temperature that does not swing with the sun.

I am not alone.

The body I touch is coiled against the left wall of the chamber. Large. Thick. Its scales are dry and cool, but the core heat coming off it is steady. My tongue reads it: snake, male, much bigger than me. His scent is old in this space, settled in. He has been here longer than the cold.

Beyond him, another. And another. The chamber extends deeper than I can see, and the floor is covered with them. Coiled shapes, pressed together, stacked in some places several deep. Their combined heat raises the chamber's temperature above what any single body could achieve on its own. The air is thick with musk and the slow exhalation of lungs barely working.

I find a space along the far wall where a smaller body is coiled against the stone. I press in beside it. Our scales touch. Neither of us moves away. There is no aggression here. No territory. The cold has taken that. What remains is the simplest survival a body can make: more mass holds more heat. I press closer. The other body presses back. We coil together not because we choose to but because the cold makes us.

The lights and darks pass. I do not count them; I cannot count them. Life in the den is not measured by sun and dark but by temperature. The stone warms slightly when the sun hits the south face above us. It cools. The bodies in the chamber shift with these changes, moving slightly when the warmth comes, settling when it goes. No one hunts. No one eats. The hunger is gone entirely now, replaced by a stillness so deep it feels like the stone itself has entered my muscles, my blood, my bones.

My heart slows. I can feel it, or I think I can feel it. The pulse that was quick and constant in the hunting is now a distant thud that comes at intervals I cannot track. My tongue barely moves. When I do flick it, the information comes back muted: same musk, same stone, same bodies, same temperature. Nothing changes. Nothing needs to change.

Sometimes the sun warms the entrance enough that a body near the top of the pile stirs and slides toward the opening. I feel the movement as a vibration passed from scale to scale through the stacked bodies. Sometimes that body comes back. Sometimes it does not. The ones that return carry the faint scent of outside air, cold and sharp, before tucking back into the mass.

I do this when the stone above is warm enough that I can feel it through layers of rock. I uncoil from the wall and work my way over and between the other bodies toward the entrance. They do not react. Some are so still that the only sign they are alive is the faintest trace of exhaled air from their nostrils, a chemical whisper that my tongue reads as living, barely.

The entrance crack is bright with the sun. I push my head out into the air, and the cold hits me like a wall. The ground outside is dry, and the sky is a hard, dry blue. The sun is low and warm on the rock face, but the air is cold, and my body starts losing heat immediately. I bask on the flat stone beside the entrance, my body stretched toward the sun, absorbing its warmth. A lizard scent crosses the rock, fresh, but I do not react. The hunger is not there. The need is shut down.

The cold drives me back inside before the sun moves a little across the sky. I slide back into the crack, and the warmth of the den takes me in, and I find my space against the wall and coil, and the small body beside me presses close again, and we settle.

This is brumation. I do not know the word as much as I know the weight of it. The way the desert contracts to a single chamber and the body contracts to a single function: hold heat, hold still, hold on until the ground warms again. There is no dreaming. There is no thinking. There is the stone and the bodies and the slow pulse and the long wait.

The cold deepens. There are stretches when the sun does not warm the entrance at all, and the den temperature drops, and every body in the chamber pulls tighter, and the breathing slows to almost nothing. I can feel the bodies around me cooling further, and the space between alive and not alive gets narrower. The pulse in my own body is so slow I cannot tell if it is still there or if I am feeling the pulse of the snake beside me or the faint rhythm of the earth itself.

Then, the stone above is warm before the heat returns.

Not much. A slight increase, maybe a little more. But the bodies in the chamber respond. A stirring runs through the pile like a vibration through a struck rock. Heads lift. Tongues move, tasting air that carries a trace of something that was not there before: moisture. Warmth. The faint green chemical signal of creosote leaves unfurling in new heat.

The long cold is ending.

I uncoil from the wall. My muscles are stiff and thin. I have burned through the supply of stored fuel, and the body that slides toward the entrance is lighter and slower than the body that entered. My scales are dull. My eyes are cloudy with the beginning of a new shed. The rattle segments click against the polished stone floor as I move, and the sound is dry and flat.

The entrance crack is warmer than the last time I passed through it. The sun outside is higher than it was in the deep cold. The ground is still cool in the shadows, but the exposed rock is warming fast, pulling heat from a sun that has shifted its angle back toward something I recognize.

I slide out onto the south facing stone and stretch my body across its surface. The heat enters me the way water enters dry ground: slowly, then faster as the channels open. My muscles loosen. My tongue starts working at full speed, reading the air thick with new scent: rain somewhere in the distance, green plants pushing through the caliche, the fresh trails of rodents that have been active through the cold I spent under the stone.

The hunger comes back.

It rises slowly from wherever it went during the long, cold, a familiar tightness behind the eyes, a readiness in the coil, a sharpening of every sense aimed at the singularity of hunting.

I am thinner. I am slower. I am alive.

The desert has done its worst, and I am still on the rock, warming in the new sun, tasting the air for anything that moves and bleeds and can be struck and swallowed.

CHAPTER SIX

BIG ANIMALS

I meet the two-leggers after the denning.

I do not know what a year means. I know desert ways of heat and cold. My body has shed several times since the creosote root. The rattle now has enough rough, hard segments to send its warning sound rolling over the pebbles. Not far, not the full dry buzz it will become. But far enough to startle a lizard several cool shaded stretches away. Far enough that the creatures rustling in the wash jerk and scatter when the rattle's sharp tremor reaches them.

I hunt along a water place, a low depression lined with something hard and smooth, neither stone nor earth. Water stays here after rain, long after the ground drinks everything else dry. The water draws things. Kangaroo rats come when the heat fades. Jackrabbits return with the heat. Javelina arrive in groups. Their hooves hit the ground in patterns I can read: heavy, irregular, and smelling of musk and half digested prickly pear. The tall four-leggers, the slow ones, also come. Their vibration starts first, a deep trembling that builds into heavy impacts, many bodies at once, shaking the earth as if the wash were running. Their heat is enormous: dense, warm masses blocking the thermal field in all directions. Their smell buries everything, fermented grass, wet dung, and urine that marks the caliche and overwrites every trail. When the tall, slow ones are at the water place, rodents hide, and hunting stops. When the tall, slow ones leave, the ground still smells of them. The rodents return, and that is when I hunt.

The water also draws birds. Birds draw me away from the edge because the sky above open water is too exposed. I hunt the margins, the mesquite and creosote ringed around the tank's overflow, where the soil is deeper, and rodent burrows are thick.

I am coiled at the base of a mesquite, fitting into a root's curve where the bark has peeled, and the exposed wood is warmed by the sun. A kangaroo rat trail runs past me. Fresh scent, grain fed, fat, careless. I wait for it to come back along its path, which they almost always do. Creatures of habit, they run the same lines between burrow, food, and water until something interrupts the circuit.

The ground changes.

A vibration comes from the east, unlike anything I have catalogued. Not the light fast tapping of a roadrunner. Not the heavy irregular shuffle of javelina. Not the scattered impacts of a jackrabbit. This is a steady rhythm: heavy, paired, regular. Impacts repeat in alternation, each landing hard and flat on the packed ground. Between the paired impacts, a pause. Then more. The ground reads it as something large walking upright.

I have never felt this pattern before.

The vibration grows. Whatever makes it comes directly toward the mesquite where I am coiled. The scent comes first: a complex chemical signature like no animal I know. It slices sharp and salty, sweat, but not javelina. Not the dusty, oily musk of coyote. Not a lizard's sun baked minerality. This is layered. Chemicals I have no reference for, sweet and acrid.

Mixed with something that smells like crushed plants, but is not.

I freeze, pressing my body low into the curve of the root. My skin flattens against the bark, blending me into the dirt and leaves. I keep my rattle hidden, making no sound, every muscle tense and ready to spring.

The paired impacts stop.

The scent presses in, pungent and close, directly above, slightly to my right. I sense its heat: not the sharp, pinpoint warmth of a rodent, nor the quick, flickering heat of a lizard. This is a tall, large thermal column radiating in all directions. The hottest point hangs high above the ground. Lower portions are blocked by something hard and dense; my pits read it as warm, but not bare skin. Hard skins on its feet. Some kind of covering below to retain heat and muffle the thermal signature.

It stands there.

A sound comes from it, from the high part, the part I cannot see without raising my head. A rhythmic vibration in the air, not the ground. Voiced. Tonal. Repeated. It is making sounds like a bird, but lower and longer, with a pattern that changes and repeats, then changes again. I cannot read it. It carries no information I can use.

The hard skins shifts closer. The ground vibration comes through my belly. It is close enough that I feel each impact separately. Another step. The thermal column moves, and the scent intensifies. Another scent arrives beneath it, sharp, volatile, faintly burning my tongue, something I will find again near the two-leggers' places.

The hard skins stops a body length from the root, where I am coiled.

I feel the air change above me. Something is coming down. Not the fast drop of a hawk. Slower. A piece of the two-legger extends downward, reaching, its heat signature narrowing to a point that moves toward the ground near my position.

I rattle.

The sound comes out of me as the strike does: not chosen, not aimed, only released. Tail segments vibrate, buzzing dry and hard between me and the reaching heat. It carries the only message I have for anything this close: I am here. Back away. This is not worth the cost.

The reaching heat stops.

The voiced sounds above change faster and higher. The paired skins step back, then again. The reaching heat withdraws upward. The thermal column straightens. For a moment, nothing moves. I rattle again, longer, louder. The coil is tight and loaded. Fangs are ready, venom glands pressurized, every part of the strike aimed at the hard skins at the ground in front of me.

The two-legger steps back again. And again. The vibrations grow lighter and farther apart, and the scent begins to thin, and the thermal column moves away, not fast, not running the way a lizard runs, but steadily, the paired impacts retreating the way they arrived.

I rattle until the vibration fades from the ground.

I rattle until the scent thins to a trace.

I rattle long after nothing is left to rattle at. My tail buzzes against the mesquite root, sound pouring out in a stream I do not try to stop. It runs itself down. The segments slow, the buzz drops to a click, and the click fades to silence.

I hold the coil long after. My body remains tense, muscles coiled and still, slow to release into the emptiness around me. The two-legger's scent remains on the ground, pressed into the dirt by those hard, flat skins. My tongue reads it rapidly. It confirms: large, strange, retreating, gone.

Gone.

But not like a hawk, climbing into the sky I cannot reach.

This thing walked away on the ground, and lives on the ground as I do. It breathed the same air. It stood at the same level and reached down toward me as I reached toward prey. It was larger than anything that had been close before, did not flinch when I froze, and did not leave until I used the rattle.

The rattle worked.

The sound useless against the sky was enough for this. The two-legger heard it, stopped, and left. My body stores this as it stores the hawk introduction, the hunting sequence, the knowledge of warmth, cold, and cover. Not as thought, but as response: two-legger vibration, heavy and paired, means coil, rattle, hold, and wait for it to leave.

I stay at the mesquite base until the ground cools, the sun drops, and the kangaroo rat trail goes stale. I do not hunt. The two-legger's scent contaminates the ground in every direction. Every rodent will smell it and avoid it.

I move west along the wash in the dark, following the bank, staying under cover, tasting the air for the smell of those hard skins and that strange layered sweat.

The desert has a new thing in it. Large, warm, walking on the ground, making sounds I cannot read, reaching toward me with something that is not a talon and not a fang and not a striking tool.

The rattle stopped it.

I hold that in my body like venom: ready, pressurized, waiting for the next. When the ground shakes with that heavy paired rhythm and the air fills with that chemical I cannot name.

CHAPTER SEVEN

BLIND

The eyes go.

I have shed many times now. The tightness, the film, the muted vision, the scraping against rough stone. The relief when the old skin peels away, and the desert sharpens. I know the sequence. My body knows better. When tightness comes, I find cover, wait, shed, and move on. The process takes what it takes. The vulnerability is brief.

This time, the process breaks.

The tightness begins as always. The scales on my flanks feel too small for my swollen body. A glossy film slides over my eyes; shapes blur, edges bleeding into a haze. I press into a crevice between flat stones on the south side of the wash, deep enough for shelter, edges sharp and rough for scraping. I have shed here before. The stone scours as it should.

I wait for the loosening. The point where the old skin separates from the new, where the edges along my lips and nostrils begin to lift, where the whole sheath starts to give, and the scraping can begin.

The loosening does not come.

The film on my eyes thickens into something denser. Not

gray. White. The desert disappears behind it and does not come back. I can see light and dark and nothing between them. A shadow crossing the sun is a change in brightness and nothing more. The creosote bush, a few body lengths from my crevice, is gone. The ground past the edge of my stones is gone. My own body, coiled in front of me, is a dim shape I know is mine only because I can feel it.

The skin stays tight. Longer than it should. Several darks and heats, too many missed hunts. The old skin is sealed to the new like wet skin dried on stone. The edges along my nose will not lift. The scales on my sides will not separate. I scrape against the rock, and the old skin stretches but does not break. I scrape harder, and the friction burns against the new surface beneath, but still the old skin holds, fused at the eyecaps, fused along the chin, refusing to let go.

The eyecaps are the worst part. They are the old skin's covers over my eyes, thin and transparent when they come off clean, but when they stay on, they thicken and harden, and the desert behind them is a white blank that my eyes cannot penetrate. I push my face against the stone, pressing the eye area against the roughest surface I can find, trying to catch the edge of the cap and roll it back. It holds. The cap has dried onto the new scale beneath it, and no amount of external pressure is breaking the seal.

I am blind.

Not the reduced vision of a normal shed. Blind. The pits on my face still read thermal contrast. My tongue still reads the air. The vibration sensors along my jaw and belly still map the ground. But the thing that tells me shape and distance and edge and movement, the thing that warned me about the hawk's shadow and tracked the gecko's path and read the two-legger's reaching appendage, that thing is gone.

I stop moving.

The crevice is safe. Nothing larger than me can enter it

from above. The opening is narrow and faces the wash, which is dry and open and carries vibration well. If something approaches on the ground, I will feel it long before it arrives. If something comes from the air, the rock above me is close, solid, and between me and the sky.

I coil in the deepest part of the crevice, and I wait.

The waiting is different during brumation. The body shuts down, and the waiting is a kind of absence, because nothing in the body moves. This is not that. I am awake. I am alert. The hunger is present, building slowly in the background. My tongue is working constantly, tasting the air, pulling in scent information that my eyes cannot confirm. A lizard passes the mouth of the crevice in the new heat. I taste it clearly: side blotched, male, moving fast across warm stone. My pits catch the faint heat of its body against the cooler rock. Every part of me that hunts says strike. But I cannot see it. I cannot judge the distance. I cannot track its movement precisely enough to gauge the launch.

I let it go. The hunger notes the loss and tightens.

A scorpion enters the crevice in the dark. I feel it through the stone before I taste it: tiny, precise impacts, eight legs tapping in rapid sequence, moving along the wall toward the back where I am coiled. Its body heat is negligible, barely above the rock. I track it by vibration alone as it moves past my coil, half a body length from my scales, and out through a crack in the back wall. I do not strike at it. I do not eat scorpions. But the fact that it passed that close without my seeing it tells me how deep the blindness goes.

A short while after, I try to shed again.

I push out of the crevice into the open air. The sun is there. I can feel it on my scales, bright and heavy, and the ground beneath me is hot. I move along the base of the wash bank, pressing my face against every rough surface I can find. A broken edge of caliche. The peeling bark of a mesquite root. The sharp rim of a stone half buried in the dirt. I drag the side of my head across each, pushing hard, trying to catch the edge of the eyecap or the lip of the old skin along my jaw.

The mesquite bark catches something. I feel a tiny tear at the corner of my mouth, the barest separation of old skin from new. I press harder. The tear extends along my upper lip, a thin line of release that runs from the corner of my mouth toward my nostril. I hook the torn edge against a knob of bark and pull forward, and the old skin begins to peel.

It comes off in pieces. Not the clean single sheath of a good shed. Strips and patches, tearing where the old skin has dried too tightly, leaving fragments stuck to the new surface. I scrape and pull and press, and the skin comes away from my head, my neck, my forward body, each piece taking effort, each piece leaving the new skin raw and oversensitive beneath it.

The eyecaps.

The left tears free when the skin around it finally separates. It comes off with a flap of old scale from my cheek, and the desert on that side rushes back: blurred, then sharpening as the new cap beneath it clears. Light. Shape. Edge. The wash bank, the mesquite, the stones, all of it returned in a short flood of information that makes the right eye's continued blindness seem worse.

The right eyecap holds. It is fused at the lower edge, dried and sealed, and no amount of scraping against bark or stone breaks the bond. I can feel the edge of it with my tongue when I flick past my own face. It is there, a thin, hard disc pressed flat against the eye, and behind it, the right eye sees nothing but white.

I shed the rest of my body skin by scraping along the wash bank, leaving strips of old skin on rocks, roots, and rough ground. The new skin beneath is bright and sensitive, and it registers more intensely than I am used to. But the shed is ugly and incomplete. The old skin did not come off smoothly, and the fragments I left behind are scattered across many body lengths of the wash bank like the leavings of something that was eaten.

The right eyecap comes off on its own. I wake in my crevice, and the sight is there on both sides, full and sharp. The cap has fallen while I was still, loosened by the moisture of my own breathing or by some final give in the bond between old and new. I find it on the stone beside my head, a tiny translucent disc no bigger than a mesquite seed. It means nothing. It is what kept me blind, and it is smaller than the smallest thing I eat.

I leave the crevice. The desert is the same. The wash, the bank, the stones, the creosote, the sky with its hawks and its hard blue weight. Nothing in it changed while I was unable to see it. The lizards are still there. The kangaroo rat trails are still fresh in the dust. The sun still works the same way on the same ground.

But I am thinner. Many suns and darks without a kill, without movement, without anything but waiting in the dark behind my own eyes. The hunger is loud and sharp, and my body is lighter than it should be, and the muscles in the strike coil feel slack.

I find a hunting position at the base of a rock near a fresh rodent trail. I coil. I wait. The eyes are clear, and the pits are working, and the tongue is reading the air, and everything is back to what it should be.

The shed left its mark. Not on my skin, which is new and bright and clean, where it came off right. I had lost ground. Many lights and darks of blindness in a desert that does not pause for anything, that kept running its rodents and its hawks while I sat in a crack in the rock and could not see my own coils.

The body remembers this the way it remembers the hawk.

Not as a thought. As a preparation. When the tightness comes again, I will find water. I will find rough stone. I will find a place where the air is moist enough that the old skin does not dry and seal before it can be shed.

The eyes are too important to lose again.

CHAPTER EIGHT

SCENT

Something is different in the air, and I do not know what it is.

It comes in the warming, after the long cold, after the emergence, after enough hunting that the body refills and the muscles rebuild and the hunger quiets to a manageable rhythm. I have lived several brumations now. I am as long as the mesquite root where I took my first gecko, and half again as thick. My diamonds are sharp on the new skin. My rattle has enough segments to fill the space between rocks with sound. I am fast and accurate, and the desert between the wash and the water tank is mine in the way that heat is mine and hunger is mine: not owned, not defended with boundaries, but occupied. I am the thing in it that other things account for.

The scent arrives on a warm dark edge when the ground is still releasing its heat and the air above it is layered with thermals that carry information in stacked ribbons. I am moving along the wash bank toward a hunting position I have used, tongue working the air in the automatic rhythm of travel, reading the usual catalogue: creosote, dust, old rodent trails, the distant tang of javelina musk from the direction of the water tank, the mineral smell of dry caliche.

Then something else.

It is not a smell I can place against anything I have encountered. It is not prey. It is not a predator. It is not a plant, mineral, water, or decay. It is biological, heavy, and it carries a chemical signature that hits the receptors on my tongue with a weight that stops me mid stride. My body reacts before my senses finish processing: the muscles along my spine tighten, the tongue flick rate doubles, the head lifts slightly from the ground, and turns toward the direction the scent is strongest.

South. Along the wash. Close to the ground, embedded in the dust, layered over the older scents the way the two-legger's sweat was layered over the dirt near the water tank. But this is not a two-legger. This is a snake.

I know the smell of other snakes. I spent the long, cold pressed against them, breathing their musk. I have crossed the trails of snakes in the wash, and on the open ground, the information was simple: snake, size, direction, and how long ago it passed. This is different. This trail is not just musk. It carries something extra, something underneath the familiar signature, a chemical I have never tasted and that my tongue is pulling in with an urgency I did not ask for.

I follow it.

This is not hunting behavior. When I hunt, I find a position, and I wait. I do not follow trails across open ground. I do not leave cover. I do not move steadily in a direction, ignoring the lizard scents and rodent trails that cross my path. But my body is doing all of these things. What drives it is not hunger. It is something else, located lower in the body, something that responds to this chemical the way the strike responds to close heat: automatically, fully, without consultation.

The trail runs south along the wash for a distance I cannot measure. The scent strengthens as I follow it. Wherever the snake that left it paused, the concentration is thicker, pooled in the dust, and I slow at each of these spots and taste them thoroughly, my tongue pressing the ground, the roof of my mouth processing the chemical in a way that feels different from how it processes prey scent or predator scent. The information is dense, and I cannot read it all. Size, yes: large, close to my own length. Direction: south and west, moving slowly. And something else. Something my body understands that has no equivalent in hunting, fleeing, or sheltering.

I follow the trail through the darkness. The ground cools, and my body slows but does not stop. The chemical pulls me forward through the cold, the way hunger never has. I cross open ground in the dark, exposed, my pattern invisible against caliche that I cannot see, and that nothing hunting me can see either. The stars are out. The air is dry and still, and the scent trail is so strong in the cool air that my tongue barely needs to touch the ground to read it.

Before the heat returns, I find him.

He is coiled at the base of a large rock on the far side of the wash, where the bank rises, and the mesquite grows thick. His body heat registers on my pits: a dense warm shape against the cooling stone, large, still. Then the scent hits full strength, not the trail anymore but the source, and the chemical that has been pulling me across the desert is suddenly everywhere, pouring off his body in waves that saturate the air between us.

He is big. Longer than me and thicker through the middle. His rattle is intact and long, with more segments than mine. His diamonds are pale in the heat, and his head is raised slightly, tongue working, reading me the way I am reading him.

We do not move toward each other quickly. I stop where his scent reaches and hold. He holds. The tongues work the air between us, and the chemical exchange is mutual and dense, carrying information I have no framework for except that my body knows what to do with it, even if the rest of me does not.

He moves. A slow uncoiling from the rock base, his body extending toward me along the ground, his chin low, his tongue touching the earth where my scent trail ends. He reaches the place where I am coiled, and his body slides alongside mine, and the contact is a shock of heat and pressure and chemical signal that runs through my scales like a vibration.

There is no dance. I have heard the combat dances through the ground, the heavy rhythmic wrestling of males pushing against each other, their bodies raised and twisting, each trying to press the other flat. I felt that through the wash bank, the vibration carried for many body lengths, and I stayed away. That was between males and had nothing to do with me.

This is different. His body moves along mine, and the pressure is steady and deliberate, and his tail finds my tail, and his chin rests on my back near my head, and the weight of him presses me into the ground in a way that is not threatening and not comfortable and not anything I have a comparison for.

The contact shifts. His tail finds mine and the lower bodies align, and what happens next is not felt on the surface. It is internal: pressure, alignment, a structural lock between his body and mine that the chemical brought both of us here to complete. My pit organs read his heat at a distance of nothing. The pheromone is so dense at this range that it crowds out the creosote, the dust, the kangaroo rat trails, everything else my tongue has been reading all night. There is only this chemical, and the pressure inside my body, and the heat of him registered on my face as a single solid warmth.

His body tenses against mine. My body answers. Something passes between us that is not venom, not heat, not scent.

Then he uncouples. His body slides away from mine, the pressure releasing, the heat of him withdrawing. His tongue flicks the air once more, reading whatever information the act has left there, and then he moves. North along the wash bank. Not fast. Not slow. The way a snake moves when it is going somewhere it has already decided to go.

He does not look back. Snakes do not look back. The head points in the direction of travel, and the body follows, and the thing that is behind is behind.

I stay at the base of the rock. His scent is heavy on my scales and in the dust around me, and the chemical that pulled me south is quieter now, not gone but changed, as if the act of arriving has altered what the signal means. My body feels different. Not injured, not tired, not the post hunt heaviness of a full belly. Something else. A shift in the internal balance, subtle, deep, located in the lower body where the contact happened.

The heat returns. I bask on the rock where he was coiled, and his residual scent mixes with mine on the warm stone. The hunger is there, quiet, waiting. I have not eaten through all of it: the trail following, the crossing, the waiting, and the body needs to eat.

I hunt. A kangaroo rat, taken at the edge of the mesquite thicket in the standard sequence: ambush, strike, release, wait, follow, swallow. The hunger eases. The body refills.

But the shift in the lower body persists. It sits there like a new weight, barely noticeable, something I carry without understanding what it is or what it will become.

The male's scent fades from my scales over the next sheds.

His trail in the wash dust is quickly gone, overwritten by wind, rodent traffic, and the passage of other bodies through the same ground.

He is gone, the way everything in the desert is gone. Completely, and without remainder.

What he left inside me is not gone. It is just beginning.

CHAPTER NINE

BURDEN

The weight builds slowly.

After the male leaves, the change in my lower body is quiet. A new density appears. There is a fullness, not food, not the lump of swallowed prey, but something else. It presses outward against my core muscles from the inside.

I hunt. I eat. The desert moves through its warming with more heat before the dark returns, hotter ground, the creosote pushing out new growth that sharpens the air with resin. I take a kangaroo rat. Later, a side blotched lizard. The hunger comes and goes in its usual rhythm, but it is shorter now. The kills do not quiet it the way they used to. My body is burning fuel faster than it should for the amount of movement I am making, as if something inside me is drawing from the same supply and I am feeding many systems instead of just me.

By the deep heat, the weight is no longer subtle. My lower body is visibly thicker. The coils that used to stack neatly now press against each other with new resistance. When I move, the added mass drags against the caliche. Crossing open ground gets harder with each crossing.

I bask more. The growing weight inside me craves heat as urgently as I do, and I feel the need in my body, a steady pull toward warm sunlit stone, warm ground, the south facing rock where the sunlight lingers. I stretch on a flat slab, my swollen lower body pressed against its warmth, soaking up the warmth, fueling the internal process with heat.

My hunting changes. The ambush positions I once used are harder to reach. I get caught in rock gaps that I slid through before. The mesquite root I coiled under now presses my distended belly; the pressure is wrong, compressing whatever is growing inside. I find wider spaces. Open ground at flat rocks. Shallow caliche depressions where my lower body can rest without compression.

My strike suffers. Not the speed, not yet, but the setup. The coil for the strike depends on a tight S-shape and even mass. The weight in my lower body throws my balance. I coil higher, stacking more of my forward body, but the angle is different. The aim shifts. I miss a kangaroo rat. It bolts into its burrow, and I taste its scent. Hunger tightens. The weight inside me does not care.

I stop eating.

This is not a decision. The hunger quiets on its own, as before brumation. The urge to hunt fades slowly; it is redirected. Everything that once fueled hunting now aims inward, at the weight, at whatever it is becoming. I do not need food. I need heat, stillness, and someplace to wait.

Clouds build. I taste the thickening moisture in the air before I see the clouds, a heaviness riding the thermals, carrying the scent of distant water. The ground is dry, but the sky shifts, the light going dense and yellow, pressure dropping in ways I feel throughout my body.

I am enormous, but not longer, no growth since before the male. I am thicker, heavier, my lower body distended. When I coil, the mass shifts and resettles. I feel the individual shapes inside me, small and dense, pressing against me and each other. They move constantly. They are alive. I do not know what they are, just that their heat moves, their pulses faint and different from my own.

I cannot hunt. I can barely move. It takes as long to cross from basking rock to the sleeping shelter. I rest halfway. I rest at the end. I rest more than I move. Resting is heavy and uncomfortable because the weight is never evenly distributed. Every surface presses, and something presses back.

The vulnerability is constant. I am so slow that a fast predator could catch me before I reach cover. I am too thick for my usual crevices. I spend the dark in wide gaps, half exposed. My rattle is ready. My strike coil is loaded as well as it can be.

Coyote passes, its footfalls quick on the far wash. I rattle from my gap. The footfalls pause, redirect, then fade. The rattle still works. It is my only full defense now.

Roadrunner finds me basking on the flat rock. It stops at a safe distance and watches, its head tilted. I rattle. It does not leave immediately, as most things do. It watches, its body still, its eyes on my distended midsection. Then it moves on, fast, its feet drumming the caliche in a quick receding pattern.

It was measuring me. Testing whether I was slow enough to take. The rattle changed its assessment, but the read was already made. That tells me what I have become: a target. Too heavy to flee, too slow to strike well, too large for my cover. I carry a weight that is not food. It will not digest or leave until it chooses.

The rains come. The first storm hits the wash with water I feel through the ground before the thunder. I am on the far bank, under a rock ledge. The water rises, brown and fast, carrying branches and debris. The scent of torn things rides the current. The air cools and wets. The temperature drops.

My body tightens around the weight inside, holding warmth, defending against the sudden chill.

The rain falls longer than usual. I hold still under the ledge. The weight inside me moves, shifting, pressing. The movements are stronger and more insistent, as if the storm's pressure is matched by an internal pressure building toward something.

The rain stops. The wash drains. The ground steams. The weight presses down.

I am not ready for what comes next, but my body is. My lower muscles begin to work without my direction. The contractions start deep and push outward. It is a new pressure, unlike the feeling of prey moving through me or the buildup to a shed. This is structural. My body opens from the inside. The weight becomes something outside of me.

It hurts. Or it is something that takes the place of pain. A massive, sustained effort pulses through every muscle in my lower body: a deep, fiery ache that does not stop, does not ease, but surges and tightens, building until the leader comes through.

Small. Wet. Hot against the cooled ground. A narrow body, slick, with a button on its tail that taps the stone once as it slides free.

Another. And another. The contractions continue. The small bodies come, each wet, alive. Each hits the ground, starts moving immediately, as I did in the packrat den, under the same rain. Small, hungry, already tasting the air.

But that is not this telling. That is the next.

The last contraction empties me. The weight is gone. My body feels lighter in every muscle, scale, and coil. The lower part that was heavy and wrong is now loose, thin, and mine again.

I am hollow, hungry, thin in new places. The energy spent supporting the weight has drawn down my reserves. What remains is a body that needs to eat, heal, and rebuild before the cold returns soon.

Small forms cluster around me on the slick stone. Their warmth barely lingers. Their scent is nearly my own, with a trace of the male, gone, with no trail left to follow.

I taste the air above them. I hold still among them.

Then the next begins.

CHAPTER TEN

LITTER

There are several of them.

They lie on the wet stone, wet, soft, moving slowly as I did inside the burrow. Each is the length of my strike. Each has a tail button that clicks faintly when touched by another or the ground. Each is already tasting the air, small tongues moving in a rhythm mirroring mine, pulling in information they cannot use yet.

Their heat is faint. Not the dense mass of an adult, but a thin warmth, barely above the stone's. My pits read them as small warm spots around my own greater heat. When they move away, their warmth fades quickly.

I do not gather it back.

The afterbirth is still on the ground. Its smell is heavy: blood, fluid, the chemical signature of the internal process that built these bodies. The smell will draw things. Ants, then beetles, then the larger scavengers that follow the scent of birth and death across the desert floor. I need to move. They need to move. But for now, we are all still on the wet stone in the last heat, and the small bodies press against my coils, and

I hold the position because my body has not yet released whatever it is that keeps me here.

A few of them crawl across my tail. I feel its belly scales on mine, the tiny grip of new skin on old, and its button taps my rattle segments as it passes. The sound it makes is nothing. A click. My rattle answers with a faint, involuntary buzz, the segments vibrating once before settling. The small body does not react. It keeps moving, following some directive that I recognize because I followed the same: away from the center, toward the edge, toward the air.

Another finds my head. It pushes against my jaw with its nose, tasting me, and I taste it back. The scent is close to mine. The undertone of the male is there, faint, buried under the stronger signature of my own body . It is slightly larger than the others. Its movements are more purposeful. It pushes past my head and moves toward the edge, where the stone drops away to the wash bed below.

I do not stop it.

The light changes. The sun is setting behind the ridge, the stone and the air are cooling, and the small bodies are losing heat faster than they can produce it. They begin to cluster. Not toward me, not specifically, but toward each other, pulling into a loose knot of overlapping coils the way my litter did in the packrat burrow. The combined warmth of small bodies raises the temperature of the cluster above what any of them can hold alone.

I am beside them but not part of the cluster. My body is coiled at the ledge, my head oriented toward the wash, my tongue reading the air for the things that the afterbirth smell will bring. A beetle is already approaching from the south, its tiny footfalls registering in the stone. A kangaroo rat trail crosses the wash bank, vast body lengths east, fresh, the rodent moving toward the water smell left by the storm in the low places. The desert is resuming its normal traffic, and the small bodies on the ledge are part of that traffic now, part of the inventory of things that move and bleed and can be eaten.

The dark comes. The stone cools. The cluster tightens. I can feel their collective heat diminishing as the temperature drops, each small body burning through its birth reserves, the fuel they were born with being spent on the simple act of staying warm through their initial dark.

I stay through the darkness. My body holds the position, and my rattle is ready, and when a larger vibration comes through the ground, something heavy and four-legged moving along the wash in the dark, I buzz once, and the vibration redirects and fades. The small bodies in the cluster do not react. They do not yet know what the rattle means. They do not yet know what anything means.

The heat returns. The stone warms. The cluster loosens. The small bodies begin to move outward, each following its own tongue, its own hunger, its own exploration of the surface it was born onto. A few of them reach the edge of the ledge and go over, sliding down the short drop to the wash bank below. I taste the air after them. Their scent trail points south along the bank, toward the creosote, the cactus, and the multitude of small shelters that the desert offers to things this size.

They do not come back.

Another moves north. Another east. They are dispersing like water disperses when it hits flat ground, spreading in every direction, following gravity and slope, and maintaining the random orientation of their bodies when they started moving. There is no pattern. I have given them no direction. They go where they go because their tongues taste something, their bellies feel a temperature difference, or their bodies simply point that way as the muscles start to push.

When the heat is highest, half of them are gone. Their scent trails fan out from the ledge in every direction. I can read each trail clearly: one went south and paused under a rock, another went down to the wash and crossed the wet sand, others followed the bank's base toward the mesquite thicket. Each trail is already cooling, the scent fading as the sun bakes the ground and the wind moves the dust.

The remaining are still on the ledge. A few are coiled together in a crack near the back wall. A single is basking on a flat stone in the sun, its body stretched long, absorbing heat. The last is at the edge, tasting the air, its button clicking against the rock as it shifts position.

I taste the air. The afterbirth smell is fading, broken down by ants, sun, and the desert's dry chemistry. The beetle is gone. The kangaroo rat trail has been crossed by a fresh set of tracks running in the opposite direction. The desert is moving at its usual speed, and the small bodies on the ledge are falling behind it.

They are venomous. I know this because I was venomous at their size, because the glands behind their small eyes are functional, because the fangs are present, and because the strike mechanism is loaded, even in a body this new. They can kill a small lizard. They can defend against a scorpion. They cannot defend against a hawk, a king snake, a coyote, or the many other things that eat small snakes in the desert. Most of them will not survive the cold. Some of them will not survive the next light and darkness.

This is not a thought I have. It is not sadness, measurement, or weighing of odds. It is nothing. They are on the ledge, and then they will not be on the ledge, and the desert will do what it does with them, the way it did with me and with my litter and with every litter born in the darkness since before the wash was a wash.

I leave.

The decision is not a decision. The thing that held me to the ledge through the dark releases sometime when the ground warmed before the sky changed, a chemical shift or a muscular relaxation or simply the hunger reasserting itself over whatever kept me still. My body uncoils from the ledge and moves toward the wash bank, and the tongue starts reading hunting information instead of birth information, and the coils find the old familiar shape of travel instead of guarding.

I do not look back. The head points in the direction of movement, and the body follows.

They are on the ledge and behind me. Their scent is on my scales and will stay there until the next shed. The male's scent, carried through them, is on me briefly and will fade.

I cross the wash and climb the far bank and follow the base of the mesquite toward the water tank. The hunger is sharp and immediate. My body is thin from carrying, and my reserves are low, and the ground is full of fresh rodent scent, and my tongue is working, and my pits are scanning, and the strike coil is rebuilding itself with every body length of forward movement.

The litter is behind me.

The desert is ahead.

I am lighter than I have been, and the lightness feels like speed, like hunger, like the body returning to the single function it does best.

I hunt.

CHAPTER ELEVEN

SHOVEL

The two-legger's den sits at the edge of my territory, where the wash flattens out, and the ground changes from caliche to something harder, darker, and smoother. I have avoided it. The smells that come from it are wrong: burning chemicals, the sharp bite of something that is not creosote, not mesquite, not anything that grows, a constant hum of vibration from inside the walls that never stops, even in the dark. The heat signature is strange, too. The walls hold warmth long after the sun goes down, radiating in a way stone does not, and the light from inside is steady and cold, carrying no warmth at all.

But the den draws rodents the way the water tank draws birds. The two-leggers leave things outside that smell of grain and fat, and the rodents come from every direction to find them. Kangaroo rat trails converge on the perimeter of the den like dry washes converging on a basin. Packrat middens cluster against the outer walls, where residual heat and food scraps create a territory richer than anything on the open desert floor.

I hunt the edges. Never close to the walls. Never on the

hard, dark ground that surrounds the den where the fast ground shakers pass and the two-leggers walk their heavy paired rhythm. I stay where the desert meets the not-desert, where the creosote gives way to cleared ground, and the rodent trails are thickest, and the hunting is easy.

I am in my apex. I have had many brumations. My body is long and thick, and the diamonds on my back are clean from a recent shed. The rattle is full voiced, a hard, dry buzz that carries far enough to turn a javelina at a distance. My strikes hit more often than they miss. The venom yield is heavy, and the kills are clean: strike, release, wait, follow the scent trail to a dead rodent within a few licks of the air. I have survived several more litters since then. I have survived another hawk and a king snake that came at me in a wash, and I drove off the hawk with a strike that connected with its midsection, but the king snake did nothing because kingsnakes do not care about venom. It left when I held my ground and rattled, and the encounter lasted longer than any encounter should, and I did not sleep that dark.

I am hunting a packrat along the edge of the two-legger's den in the warming after brumation. The trail is fresh, the scent heavy with grain and nesting material. The packrat has crossed from a midden against the den wall to a burrow entrance under a prickly pear leading into the desert. It will come back. They always come back.

I am coiled at the base of the prickly pear, my body fitting into the depression where the roots have softened the soil, my head oriented toward the den, my tongue reading the air in the easy rhythm of a set ambush. The ground is warm. The light is long and low. The packrat scent is everywhere.

The vibration comes from the den.

Paired heavy rhythm. Two-legger. But faster than usual, the impacts are harder, and accompanied by a different pattern: lighter, quicker, four-legged, low to the ground, moving ahead of the two-legger in an erratic pattern. The four-legged thing is hot and fast, and its scent arrives before its footfalls register clearly: wet breath, grain fed, the sharp chemical smell of the two-legger's den embedded in its fur. The fast, low hunter that stays close to the two-leggers and moves without purpose, and makes noise from its throat that carries across the open ground.

It reaches my position before the two-legger does.

It does not see me. It smells me. Its nose drops to the ground near the prickly pear, and the exhaled breath hits the dust in front of my face, hot and wet and reeking of unnatural meat and the chemical saliva of a grain fed carnivore. It snorts. Its body goes rigid. The noise from its throat shifts from random vocalizations of movement to a sharp, repeating bark aimed directly at my coil.

I freeze. I flatten. My pattern against the soil, the root, and the dead prickly pear pads is good. In the fading light, I am grounded.

The thing does not care about the pattern. It smells me, and the smell overrides the eyes, and it pushes closer, its nose a few body lengths from my head, its bark continuous now, high and hard and aimed at the two-legger behind it.

I rattle.

The full voice. Every segment buzzing against every other segment, the sound filling the space between me and the furry four-legger and the prickly pear and the open ground. The thing jumps back. A body length, now more. The barking does not stop, but the nose retreats, and the body holds at a distance defined by the rattle.

The two-legger arrives.

Heavy paired footfalls, running now, the rhythm broken and fast. The thermal column is large and tall, blocking the last of the western heat. Its scent is sweat and the chemical compound I learned at the water tank, and something new underneath: a sharp spike in the scent profile that I will later encounter again and again in the presence of two-leggers who have seen me. Something their bodies produce when the rattle sound reaches them. A fear chemical. But mixed with something else, something hotter, something that does not smell like retreat.

The two-legger's voiced sounds are loud and fast. The four-legger moves to the side but does not leave. The two-legger's footfalls stop body lengths from my coil.

I rattle harder. The coil is loaded. The fangs are ready.

The two-legger moves. Not away. To the side. Then back toward the den. The footfalls recede, and for a moment I think the sequence is running the way it ran at the water tank: rattle, pause, retreat. I hold the coil. I hold the rattle.

The footfalls come back. Faster. The two-legger is closer now, closer body lengths, and the thermal column has changed shape. An appendage, the upper part, is extended and holding something. A long shape, thin at one end and wide and flat at the other. The flat end is high above me, above the two-legger's own head, and the heat signature of its hands grips the thin end, and the muscles in the thermal column are tensed in a configuration I have never read before.

The flat thing starts to come down.

I am already in the escalation. Freeze happened when the four-legged arrived. Coil happened after. Rattle has been running continuously. The next step is mock strike: a fast lunge with the mouth closed that covers half the distance, then shows the fangs, and says, "This is real, this is not a sound, this is what comes next."

I mock strike. My head launches forward and snaps back, and the speed of it is the fastest thing my body can produce, and the two-legger flinches, the thermal column rocking backward, the flat thing wavering in the air above me.

But it does not leave.

The flat thing comes down.

It hits the ground beside me. Not on me. Beside me. The impact comes through the earth like a shockwave, a flat, hard shockwave that travels through the caliche and my belly scales, shaking every bone in my coil. Dirt and gravel spray against my scales. The edge of the flat thing is sharp, and it cuts a line in the soil close to my body.

I strike. For real. The mouth opens, and the fangs swing down, and the lunge carries me forward, and I feel the fangs hit something hard and smooth that is not flesh and not bone and not anything alive. The flat thing. My fangs scrape across its surface, and the venom sprays against the metal, and the strike slides off and connects with nothing.

The hard flat thing rises and comes down again.

This time it connects. The edge catches my body lower down my length, at the base of the rattle by my stripes, and the force of it is enormous, a crushing weight that pins me to the ground, and the pain is a white flash that runs from the point of impact to my skull. I feel the rattle segments shatter. Half of them, maybe more, broken free from the base and scattered on the dirt. The tail goes numb below the impact point.

I pull.

Every muscle in my body fires at once, a full length contraction aimed at getting free of the pressure that has me pinned. The flat thing lifts slightly as the two-legger adjusts for another swing and in that moment I am moving, pulling my damaged tail through the gap, dragging my body across the dirt toward the prickly pear, toward the root system, toward the gap between the pads where a body my size can go and a flat thing on a stick cannot follow.

I reach the gap.

The prickly pear spines rake my scales as I push through. The flat thing hits the ground behind me, another shockwave in the soil that I feel through my belly as I slide into the root space and coil against the base of the plant.

I rattle. What is left of the rattle. The segments that remain buzz unevenly, the sound broken and rough, missing the full, dry voice I have carried for most of my life. The tone is different. Damaged. Shorter. Crooked.

The two-legger stands outside the prickly pear. I can feel the footfalls shifting, the thermal column moving around the perimeter, looking for an angle. The four-legged creature is barking from a distance, its four-legged pattern circling. The flat thing taps the ground, probing.

I hold the coil. I hold the broken rattle. The venom glands are full. The fangs are intact. The strike is loaded.

The two-legger makes more voiced sounds. The four-legged barking fades to a whine. The footfalls shift again, pause, and then begin to retreat. Slowly. The paired heavy rhythm moves away toward the den. The four-legger follows. The scent of both of them lingers in the air long after the vibrations fade.

I hold the coil until the ground goes completely still.

The pain in my tail is a steady pulse that does not quiet. The broken segments lie on the dirt behind the prickly pear, drying in the cooling air. I can smell them. My own keratin, my own rattle, separated from my body and lying in the dust like the shed skins I have left on rocks and roots across the desert.

I stay in the prickly pear through several darks and lights. The tail swells where the flat thing hit. The muscles around the break are stiff and hot with the body's repair response. I do not hunt. I do not move. The hunger comes back, and I ignore it.

After several disappearances and reappearances of the sun, I uncoil and slide out of the prickly pear on the side facing away from the two-legger den. My movement is slower. The damaged tail drags differently, the balance altered by the missing segments. The rattle buzzes when I move across rough ground, but the sound is wrong, shorter and rougher, a damaged instrument playing a damaged song.

It still warns. I test it against a lizard that freezes at the sound. I test it against the ground vibration of a jackrabbit that redirects its path. The sound is different, but the message is the same.

"I am here. Back away. This is not worth what it will cost you."

I move west along the wash, away from the two-legger's den, away from the packrat middens and the grain scented trails and the place where the desert meets the not desert. The hunting there is good, but the hunting there almost killed me, and the body does not forget. The paired heavy footfalls, the four-legged bark, the flat thing rising and falling, and the white flash of impact.

The rattle grew back. It always grows back. Each shed adds a segment, and over the sheds, the voice rebuilds. But it rebuilds crooked, the new segments growing around the scar of the break, and the sound it makes now is not the sound it made before.

Rougher. Uneven. A crooked song.

Good. A crooked song still warns.

CHAPTER TWELVE

LEAN

The missed strikes begin after brumation and the cold.

A kangaroo rat sits at the mouth of its burrow in the early dark. My body is a warm pulse against cool ground. Not long ago, I would have taken it without thinking. The coil loads. The strike fires. My head launches forward, fangs swinging down. The distance closes. The fangs hit dirt. Gone. Not because it was faster than my other prey. Because I am slower than before.

I taste the scent trail in the dust where it sat. Fresh, warm, heavy with seeds. The trail leads to the burrow, and the burrow is dark and deep, and the rat is inside it, alive, because my strike arrived a breath too late.

I recoil and wait. The rat does not come back out. They learn fast when the strike misses. This burrow is finished for me. I will not hunt it again for a while.

Hunger tightens.

I am old. I do not know this the way a two-legger knows it, with their numbers and shiny glass and the memory of what the body used to be. I know it the way I know the temperature of the ground beneath me: through contact, through what registers in the muscles and the bones and the speed of every movement I make.

The strike is slower. Not by much. Not by a margin I could gauge in a single attempt. But the misses accumulate. It happens increasingly. Prey I would have taken cleanly is now able to react, to flinch, to close the gap between fang and flesh before the fang arrives. The launch holds. The coil loads as always. But delivery, the final instant where speed means contact, has lost an edge I can't restore.

The sheds are harder. The skin does not loosen as it did when I was young. Once tightness set in, separation followed, and the whole sheath peeled off, one clean pull against rough stone. Now, the process drags on. The film on my eyes lasts longer. The skin clings at the edges, along the jaw, around the eyecaps, tight places where old and new scale meet. The bond stays; it doesn't release cleanly. I spend more time scraping. I spend more time blind. I wedge myself into crevices, waiting for my body to finish a process it no longer performs efficiently.

The eyecaps stick a few times. I worked them free against a mesquite root. Next time, the right cap stayed sealed for a long, hungry period. When it finally came off, my vision beneath was less sharp than before, the desert blurs at the edges, shapes lose definition, and instead of a clear divide, a warm body blends faintly into the cooler ground. I compensate by hunting closer, setting ambushes where the strike distance is shorter, so the speed I have lost does not matter. The prey is already within reach when the coil fires.

The hunting territory shrinks. Not because another has pushed me out. Because the distances I used to cover in travel now take longer and cost more. I move less. I cover less ground. The wash bank I traveled end to end is now divided into sections, and I hunt a section until it is empty, then move to the next. The moving takes energy that the hunting does not always replace.

I eat less. Not from lack of hunger. The hunger is constant. It is the same pull behind the eyes that has driven me since the packrat burrow. But the kills come further apart. A kangaroo rat after too long. A lizard between the rats when I can get them. Long stretches of nothing, coiled under a rock or at the base of a bush. I wait for something warm to cross my smaller range. My body burns through reserves that aren't replenished as fast as they're used up.

I am thinner. The thick body of my youth, the heavy muscled coil that could load a strike and hold a position without fatigue, has narrowed. My scales sit closer to the ribs. The diamonds on my back, still sharp from the most recent shed, pattern a body that has less beneath them than it used to.

Yet I still hunt. I still kill. A packrat captured in the dark near a burrow mouth, clean strike, fangs meeting flesh, venom working, the wait, the trail, the swallow. The mechanism still functions. The venom glands refill. The fangs still replace, though slower now, the new teeth coming in behind the old with a delay I notice in their looseness. The jaw still unhinges. The throat still expands. Muscular swallowing still moves prey from the mouth to the belly.

But the intervals between kills are longer, and the body between kills is thinner, and the reserves that carry me through the lean times are shallower than they used to be.

The rebuilding stretches longer now. Sometimes into the full heat. The other snakes in the den are not the same from one brumation to the next. Some I recognize by scent. Others are new, young, thick with reserves I no longer carry. I coil against them. Their warmth helps, and mine helps theirs. The exchange is the same as always, but I bring less to it.

After my last brumation, I emerge into a desert that

smells the same as it always has. Creosote and dust and the distant hint of water from somewhere south. The sun is where it should be. The ground warms at the rate it always warms. The kangaroo rat trails are fresh in the dust along the wash bank.

My body does not respond the way it used to.

Warming takes longer. I bask longer on the south facing rock. The heat enters slowly. It's as if the channels that carry warmth from scales to muscles have narrowed. The muscles do eventually loosen, but they never return to what they were, full, elastic, ready. That made the strike fast, travel easy. The coil? A loaded mechanism that could fire without thought. Now there is stiffness that the sun cannot erase. A new resistance in the jaw's joints, along the spine, in the tight place where the rattle meets the body.

I hunt the closest ambush positions. The rock at the edge of the wash where the rodent trail passes within half a body length. The base of the prickly pear is where the lizards bask in the warm early light. The gap between stones where I can coil and wait, and the strike distance is short enough that the speed I have left is enough.

There are stretches when I do not hunt at all. The hunger is there, but my body does not mobilize for it as before. I bask on the rock. The sun works on my scales. The tongue reads the air. The information comes in. I process it. I do not move. A lizard crosses close enough to taste. I note it. The coil does not load. A kangaroo rat runs its trail body lengths east. I taste it. The strike mechanism does not engage.

I am conserving. Not by choice. The body is rationing what it has. It spends the minimum. It holds reserves and waits for something close and certain enough that the strike will be justified by the chance of a kill.

The desert does not slow down for me. The hawks still circle. Kingsnakes still hunt the wash. Coyotes still pass in the dark. Their quick, light footfalls vibrate through the ground beneath my coil. The two-leggers still walk the edges of my territory with their paired, heavy rhythm and chemical smell. Everything moves at the same speed it always has. I am the thing that has changed.

The rattle still works. It is shorter than it was in my peak. Crooked segments from the injury now give it a rough, uneven buzz that is mine and no other snake's. When I use it, things that hear it still stop. They still redirect. They still give me the space the sound demands. The sound is enough. It has always been, except for what refuses to listen.

I shed again in the deep heat. The process takes longer than ever. The eyecaps stick on both sides. I scrape them free against the rock at the den's mouth, where generations of snakes have polished the stone with this very movement. The new skin is bright. The vision is clear now. The desert, poised before the cold, is the same as always: vast, dry, full of things that move and bleed and can be struck and swallowed by whatever has the speed to take them.

I coil on the south facing rock. The sun is warm. The ground is warm. The air carries the beginning hint of the coming cold, a dryness and a thinning that I feel in the way the thermals drop sooner than they did.

The heat is thinning.

I will enter the den again when the cold comes. I will coil in the dark against the other bodies, and the slow pulse of shared warmth and the long nothing of brumation will take me under, and maybe I will emerge again, and maybe I will not.

The body does not think about this. The body does what it does. It finds warmth. It holds still. It waits for the next thing that comes close enough.

The rattle clicks against the rock as I shift my coils.

I wait.

CHAPTER THIRTEEN

LAST ENCOUNTER

After brumation is thin.

The ground warms late. The sun reaches the south face of the den later each light than I remember, and the heat it brings is less, and the cold that follows into the dark comes sooner. My body emerges from the crack in the rock at a moment when the stone is barely warm enough to move on, my muscles stiff, my tongue slow, the coils unwinding from the tangle of bodies in the chamber below with an effort that takes more than I can afford.

I am lighter than I have ever been after I attained my full size. The reserves that carried me into the den were low, and the long, cold underground has drawn them down to something I can feel in the way my scales sit loose against my ribs and the way my head feels heavy on a neck that does not hold it as high as it once did. The diamonds on my back are dull. The shed that should have come before the cold never completed, and patches of old skin still cling along my flanks and at the base of my jaw, dry and pale against the darker new growth beneath.

I bask on the south facing rock. The same rock. The same sun. The warmth enters me slowly, slower than before, as if the body has forgotten how to accept it or as if the channels between the scales and the blood have narrowed to threads. I stretch my full length across the stone, and it is not as long as it sounds. The muscles are thinner. When I pull the coils in, they do not stack to the density they once had.

The rattle clicks against the rock. Crooked. Short. Missing the segments the flat thing and two-legger took, regrown around the scar in an uneven line that buzzes rough when I test it. I test it now with a short burst against the stone, and the sound carries across the air, and a lizard on a nearby rock freezes, then runs. The sound still works. That is what matters.

I hunt the closest position. The rock where the kangaroo rat trail passes is within a body length of it. I coil there in the small shadows and wait. The tongue works the air. The trail is fresh, but the rat is in its burrow, and I wait for it the way I have waited for things in this desert, still and loaded and aimed at the place where the warm body will appear.

It appears. A quick shape darting from the burrow mouth to a seed cache under a creosote root. My pits read it. My eyes, clouded at the edges, track it. The coil fires.

The strike misses. Not by the narrow margin of last misses. By a distance that tells me the speed is gone in a way that is not coming back. The rat is in its cache before my fangs reach the ground where it was sitting. I taste the dust. Empty.

I recoil. The effort of the strike has cost me more than it should have. My muscles are slow to reset. The coil does not rebuild as fast. The heart, which used to drive the strike recovery in moments, is laboring in a way I feel behind my ribs, a heaviness in the rhythm that was not there before.

I waited at the rock. Nothing else comes close enough. I move back to the basking stone, and the sun is past its peak, and the warmth it gives is the diminishing warmth that has already spent its best. I absorb what is there. It is not enough.

I begin to go this way: Basking. Waiting. The occasional hunt that produces nothing. A lizard taken in the bright, warm. It was small, barely enough to register in my belly. Then a long stretch of nothing. I waited, coiled under the flat rock, the hunger constant, the body spending reserves it could not replace.

The warmth deepens into heat, and the heat burns as it always does, and the hunting does not improve. I take a kangaroo rat in the rains, a clean kill that feels like a memory of what I used to be, the strike connecting, the venom working, the trail leading to a dead body in the dirt. I swallow it, and the fullness lasts longer than it should because my body is processing it more slowly, extracting what it can at a rate that matches the diminished pace of everything else.

The next cold comes early. Or it comes, as it always does, and my body registers it sooner because there is less between me and it. The ground cools, and the muscles stiffen, and the tongue slows, and the long pull toward the den begins before the other snakes have started moving that way.

I follow the trail to the outcrop. The same trail, worn into the dust by passage. The same south facing rock, warm in the low sun. The same crack at the base, breathing the musk of bodies already gathered inside.

I slide in.

The crack is the same width it has always been, but my body moves through it differently. Slower. The muscles that used to push me through the narrowing in a single smooth pull now work in segments, contracting, resting, and contracting again. The polished stone floor is cool against my belly, and the temperature rises as I descend, and the scent of the den fills my tongue with the old familiar catalogue: snake musk, full of bodies, stone, earth, the deep still air of a place that does not change.

I found a space along the wall. The same wall. Close to the place where I spent my initial brumation, pressed against a body I did not know. There is a body here now, already coiled, already cool, its heat banked low for the long wait. I press against it. Our scales touch. The combined warmth is slight, but more than either of us alone.

The den is filled. I feel them arriving through the stone, each new body a vibration in the entrance crack, a shifting of the mass on the polished floor, a fresh layer of musk added to the air. Some are large; their heat signatures dense and steady. Some are small, young, their warmth faint and quick. They pile in, and the chamber's temperature rises incrementally as the bodies accumulate.

I am deep in the pile. Not by choice. The bodies that arrived after me pushed me farther from the entrance and closer to the back wall, where the stone is coldest, and the air is still. I do not move to correct this. The effort of repositioning, of pushing through the stacked coils and tangled bodies, is more than the difference in temperature is worth.

The cold comes. The surface cold, the deep cold that sinks through rock and soil and reaches the chamber in a slow wave, lowering the temperature. The bodies respond. The pile tightens. The breathing slows. The hearts, all of them, drop to their lowest rhythm, each a faint pulse in the dark, barely distinguishable from the pulse of the snake beside it.

My heart slows with them. The beat that has driven my blood from the packrat burrow to this moment, through sheds and strikes and the long stillness of ambush, through the hawk and the two-legger's sharp thing and the weight of carrying young and the slow failure, that beat drops to a pace I have not felt before. Slower than any brumation.

Slower than the deepest cold of the deepest cold dark. A rhythm that is not counting but releasing.

The hunger is gone. It left somewhere in the early cold and did not come back. The pull behind the eyes, my companion since the beginning moments in the packrat burrow, is quiet. Not satisfied. Not waiting. Quiet.

The tongue stops working. The last flick tasted of stone and musk, and the faint signature of the body pressed against mine, and then the tongue rested against the floor of my mouth and did not move again. The information it was gathering had not changed. The same air. The same bodies. The same temperature.

My pits read the surrounding bodies. Their heat is faint, barely above the stone, dim shapes in a dim field. Beside me is a steady, low glow. Across from my tail is dimmer. My heat is dimmer still, a reading I would have difficulty detecting in another body, a thermal signature sinking toward the rock beneath me.

The cold pressed in. Not fast. Not violent. The way the sun presses on stone. Steady. Constant. It has been pressing on this den for longer than I have been alive, and it will press on it after all inside has cooled to its temperature and been replaced by new bodies that will cool in their turn.

The muscles in my coil loosen. Not the loosening of warmth returning. A different loosening. The tension that has held the coil, the spring that loaded the strike, the readiness that kept the body aimed at the next warm thing, that tension is draining out through the belly scales into the stone, the way heat drains out in the especially cold. It does not come back when the stone warms. It does not come back.

The heart beats. And beats. And the interval between the beats stretches. And stretches.

The body beside me is warm. The stone beneath me is

cold. The air in the chamber is still and dark and full of the slow breath of living things that are almost not breathing.

The rattle, crooked and short and scarred, rests against the polished stone where as many tails as there are stones in the wash have rested before mine. It does not buzz. It does not click. The segments are still.

The dark is

APPENDIX

The Western Diamondback Rattlesnake Crotalus Atrox

The Western Diamondback Rattlesnake is named in two languages: Crotalus, from the Greek for 'rattle,' and Atrox, from the Latin for 'fierce.' Spencer Fullerton Baird and Charles Frédéric Girard described it in 1853 from American Southwest specimens. The name Atrox reflects the fear it inspired, though the snake is not aggressive. Instead, it is patient, precise, and avoids conflict.

Range and Habitat

Crotalus Atrox is widely distributed in North America, from southwestern Oklahoma and central and western Texas through New Mexico, Arizona, southern California, and into northern and central Mexico. It lives in desert scrub, grassland, rocky hillsides, and near farms, especially in the Chihuahuan and Sonoran deserts. The species adapts well to a wide range of elevations, soils, and vegetation, making it a successful large pit viper.

Physical Description

Adults usually reach three and a half to five feet, with some reaching over seven feet. The body is heavy, with dark diamond patterns outlined in lighter scales. The tail shows alternating black and white rings near the rattle, sometimes described as a raccoon tail, which is useful for identification at a distance. The head is triangular and broad, housing venom glands. Between the eye and the nostril is the loreal pit, an organ that detects infrared radiation, allowing the snake to locate warm-blooded prey in total darkness and aim a strike without using its eyes at all.

Reproduction and Birth

The Western Diamondback gives birth to live young, not eggs. Litters range from four to twenty-five, usually in late summer following rain. Newborns are fully formed, eight to thirteen inches long, and venomous. The mother provides no care, and the young disperse soon after birth.

The Rattle

The rattle is made of hollow keratin segments, like fingernails. Each skin shed adds a segment, but segments break off, so their number doesn't show age. The rattle makes a sound when the tail vibrates rapidly, signaling a warning, not aggression. It is meant to prevent conflict; the snake prefers to avoid contact.

Feeding and Venom

The Western Diamondback is an ambush predator that waits along animal trails. Its strike is almost invisible. It tracks prey by scent and swallows it whole. Its venom ranks among the most potent hemotoxic venoms in North America, capable of breaking down tissue and disrupting clotting at the cellular level, a precision instrument evolved over millions of years to immobilize prey, not to threaten the animals large enough to step around it.

Lifespan and Mortality

Western Diamondbacks face predators like hawks, eagles, roadrunners, kingsnakes, and humans, and road deaths are common. Many die in their first year; survivors can live fifteen to twenty years. Growth slows with age. Sexual maturity is reached between three and five years. The snake is ectothermic, brumates in cold months, and is most active at dawn, dusk, and night in warmer weather.

The Western Diamondback avoids conflict and does not defend territory. It is a product of its environment, shaped by evolution. Its behavior is driven by efficiency, not aggression. The snake waits because waiting works. Atrox is our name for it. The snake has its rattle, its range, and enough.

ABOUT THE AUTHOR

Cara Van Carter is a nomad living in Arizona with her husband and children. When she is not writing or observing the desert, she is making tortillas and raising her children.

For permissions, inquiries, or correspondence:
Coyote Pack Publishing
info@coyotepackpublishing.com